THE MICROPEZIDAE OF CALIFORNIA
(DIPTERA)

An adult micropezid fly.

BULLETIN OF THE CALIFORNIA INSECT SURVEY
VOLUME 14

THE MICROPEZIDAE OF CALIFORNIA
(Diptera)

BY

RICHARD W. MERRITT

(Department of Entomological Sciences, University of California, Berkeley)

and

MAURICE T. JAMES

(Department of Entomology, Washington State University, Pullman)

UNIVERSITY OF CALIFORNIA PRESS
BERKELEY   ·   LOS ANGELES   ·   LONDON
1973

BULLETIN OF THE CALIFORNIA INSECT SURVEY

VOLUME 14

Approved for publication October 29, 1971

Issued February 20, 1973

UNIVERSITY OF CALIFORNIA PRESS
BERKELEY AND LOS ANGELES

UNIVERSITY OF CALIFORNIA PRESS, LTD.
LONDON, ENGLAND

ISBN: 0-520-09435-2
LIBRARY OF CONGRESS CATALOG CARD No.: 75-185975

CONTENTS

THE MICROPEZIDAE OF CALIFORNIA (Diptera)

BY

RICHARD W. MERRITT and MAURICE T. JAMES

INTRODUCTION

THE MEMBERS OF Micropezidae, commonly known as the stilt-legged or long-legged flies, comprise a small family of acalyptrate Diptera. Approximately 90 percent of the world species are tropical and more than half are found in the New World (Cresson, 1938a). Twenty-two species occur in the Nearctic west of the Rocky Mountains, and, of these, seven are known to occur in California. However, in the present paper all the western species are included in the keys and discussion, since all of these might be considered as possible residents of California. Distributional data, in summary form, of each species for the western states other than California was presented by Merritt (1972).

Two important works on the taxonomy of the North American Micropezidae are those of Hennig (1934, 1935a) and Cresson (1938a). Other species from North America have been described subsequently by Cresson (1938b) and James (1946). The classification used in this paper is that of Steyskal (1965).

Several illustrations (figs. 1, 8–14, 35–43) were taken directly from Cresson (1930, 1938a) and two original sketches drawn by Steyskal (figs. 32 and 33) were slightly modified and, with his permission, included in the study. All other illustrations were prepared with the aid of a squared reticle and an ocular micrometer disc.

In analyzing the distribution of the principal micropezid genera found in California, two interesting trends were observed. The genus *Compsobata*, consisting of a group of Holarctic species, appears to possess affinities in California with species from the north and has apparently not been recorded much farther south than 36° latitude. Members of this genus have often been collected in marsh or wet meadow habitats and at altitudes greater than 3,000 feet. Members of the genus *Micropeza*, which is largely Neotropical in distribution, extend into California from the south and do not occur farther north in California than Mendocino County. Adults usually occur in more xerophytic areas than do those of *Compsobata* (map 1).

BIOLOGY

There is relatively little known of the biology of the Micropezidae. Oldroyd (1964) stated that the adults are predacious on other insects, such as aphids and small flies, which they stalk through the vegetation. However, our observations do not support this predation hypothesis. The mouthparts appears to be of the sponging type and not modified for predation. Cole (1969) cited Schiner as recording one species on human excrement in Europe, while Berg (1947) noted in the Solomon Islands that adults fed on excrement, carrion, and putrescent fruit. D. Viers (personal communication) found

* Scientific paper 3681, College of Agriculture, Washington State University, Pullman, Washington. Work conducted under Project 9043.

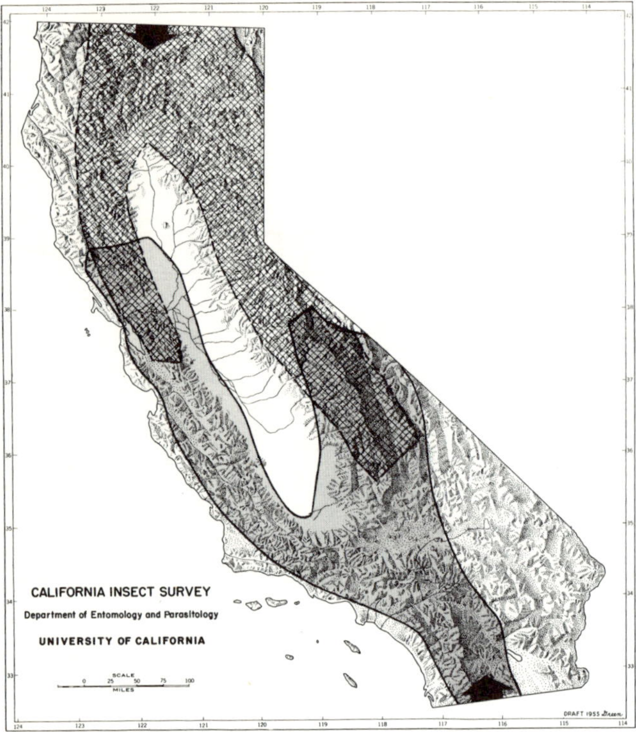

Map 1. California distribution of: ✕✕✕✕✕✕, *Compsobata* Czerny; ▓▓▓▓▓▓, *Micropeza* Meigen.

adult specimens belonging to unidentified species of *Scipopus* Enderlein and *Cardiacephala* Macquart attracted to human, primate and avian excrement in Costa Rica.

Adults have been observed running on the upper surfaces of leaves or bark of trees in the jungle underbrush, with their wings folded over their abdomen and their white fore tarsi rubbing or waving in front of them (Berg, 1947). R. D. Akre (personal communication) observed them usually on horizontal surfaces, bark, leaves, etc., exposed to sunlight, when exhibiting this display. This behavior was also observed personally by one of us (Merritt) in Costa Rica. Hennig (1935*b*) emphasized that this habit often contributes toward a remarkable mimicry of certain Ichneumonidae, and Oldroyd (1964) reported that some tropical Micropezidae are "mimetic" in the sense that they look like anything but a fly. Wheeler (1924) gives an interesting account of the courtship display of Neotropical micropezids.

In the United States, west of the Rocky Mountains, adults are usually collected in meadows, marshes, moist woods, on the bark of trees, or on stems and leaves of herbaceous plants. We have recorded the adults on *Cleome serrulata* (Rocky Mountain bee plant), *Robinia neomexicana* (locust), *Apocynum* sp. (Indian hemp), *Franseria deltoidea, Salix* sp. (willow), field corn, and

alfalfa. They have been collected at altitudes up to 11,000 feet.

Little work has been done on the immature stages of Micropezidae. It appears that the larvae generally develop in rotting wood, fruit, or other vegetable material. Cresson (1938*a*) reported the rearing of *Taeniaptera lasciva* (Fabricius) from decaying sugarcane cuttings that had failed to germinate. Sabrosky (1942) found larvae of *Rainieria brunneipes* (Say) in the crotch of a large American elm, and Berg (1947) discovered puparia of *Mimegralla albimana striatofasciata* Enderlein in the wet bark of a large hardwood tree trunk. Steyskal (1964) reported the larvae of micropezids attacking the root of ginger and in this same paper he discussed the larval morphology and biology and presented a key to the known third instars of Micropezidae. Brindle (1965) included a short key and descriptions of two species of British micropezid larvae found in a decayed grass heap. Wallace (1969) described the larva and pupa of *Calobatina geometroides* (Cresson) from Georgia, collected in detritus in a tree trunk. More recently, Teskey (1972) described the mature larva and pupa of *Compsobata univitta* (Walker) from Manitoba, found in a pile of decaying vegetation by a stream.

EXTERNAL MORPHOLOGY

Because of the unusual development of the sclerites of the micropezid head, Cresson (1938*a*) proposed several terms not generally used in Diptera taxonomy and changed the meanings of several well understood terms. To maintain stability we have accepted Cresson's definitions but have included only those applicable to this study. We have adapted the following definitions from Cresson (1938*a*), as paraphrased here (fig. 1):

The term *occiput* refers to the entire region of the head posterior to the eyes. The *paracephalon*, its lateral portion, bordering the posterior orbit and bearing the outer vertical bristles, is greatly developed in *Micropeza*. The *epicephala* are above and mesad of the paracephala and bear the inner verticals and the posterior pair of frontal bristles; these generally occupy the lateral angles of the vertex and extend posteriorly to the foramen; anteriorly they are usually not distinctly differentiated from the *frontalia*. The *frontalia* are often very narrow or may be indeterminable. Between the epicephala and the frontalia are the *mesofrons*, often an opaque, median stripe, bearing the *ocelli*, and extending more or less the entire length of the frons. Laterad of the mesofrons anteriorly is the *parafrons*, not always differentiated, generally extending posteriorly along the orbits to near the vertex. In the genus *Micro-*

peza we consider the parafrons to be the narrow ridge adjacent to each eye and delimiting the mesofrons. The frontalia are not differentiated in this genus. Other terminology should offer no problems.

The male claspers in many species are extremely characteristic and used for species determination. Unlike the periphallic organs of many other Diptera, they arise from the fourth and fifth abdominal sternites and are therefore extremely conspicuous. They may be pronglike digitate processes as in *Micropeza* (figs. 21, 23, and 25), or conical processes opposing each other as in *Compsobata* (figs. 15, 16, and 17). *Compsobata* also possesses a characteristic bilobed structure, the fulcrum (fig. 16F), arising from an area between the claspers. This structure is also used in species determination. Cole (1927) discussed the modifications in the terminal structures of Micropezidae and Crampton, et al. (1942) provided excellent illustrations and labeled the sclerites and terminalia.

Color variation is commonly encountered in the Micropezidae. Merritt (1970) found regional melanism involving leg coloration in *Compsobata mima* Hennig. A detailed account of this is given in the discussion of *C. mima* Hennig. In *Micropeza lineata* Van Duzee, the females possess black and pale antennae while the males have only pale antennae.

ACKNOWLEDGMENTS

Specimens were examined through the courtesy of the following institutions and their curators: California Academy of Sciences, San Francisco (C.A.S.); California Department of Agriculture, Sacramento (C.D.A.); California Insect Survey, University of California, Berkeley (C.I.S.); University of California, Riverside (U.C.R.); University of California, Davis (U.C.D.); University of Arizona, Tucson (U.A.); University of Kansas, Lawrence (U.K.); Utah State University, Logan (U.S.U.); Oregon State University, Corvallis (O.S.U.); Oregon Department of Agriculture, Salem (O.D.A.); American Museum of Natural History, New York (A.M.N.H.); Academy of Natural Sciences of Philadelphia (A.N.S.P.); Washington State University, Pullman (W.S.U.); and University of Idaho, Moscow (U.I.). We wish to thank Max W. McFadden for many helpful suggestions and E. S. Ross for permission to use his excellent photograph for the frontispiece.

SYSTEMATIC TREATMENT

SYNOPSIS OF THE FAMILY MICROPEZIDAE

THE MICROPEZIDAE can be distinguished from other acalyptrate Diptera by the short antenna with a subbasal arista, an elongate thorax with the wings long and slender, and stiltlike legs. The abdomen is long and narrow. The first posterior cell divides the wing longitudinally and is narrowed or closed at the margin. The second basal crossvein is united with the discal cell in *Micropeza* and separated in *Cnodacophora* and *Compsobata* (fig. 5). The large, characteristically shaped claspers arising as modifications of the fourth and fifth abdominal sternites will, alone, distinguish many male Micropezidae from other families. The phallic organs are small and the ovipositor is large and pendulous.

Steyskal (1964) presents the following excellent summary of the taxonomic status of the family Micropezidae: "Hennig in 1952 considered the Neriidae a subfamily of the Micropezidae (Tylidae), but in his more recent work on the families of the Diptera Schizophora (Hennig, 1958) he includes five families in the superfamily Micropezoidea *sens. lat.*: Micropezidae, Taeniapteridae, Calobatidae (Trepidariidae), Neriidae, and Cypselosomatidae. Removal of the last-mentioned family, which consists of two rare oriental species, leaves the superfamily Micropezoidea *sens. str.*, a group usually considered a single family, Micropezidae."

As noted above, the Neriidae was at one time considered part of the Micropezidae. Often these two families are still confused but may be separated by the following characters: third antennal segment decumbent in the Micropezidae as opposed to porrect in Neriidae, second antennal segment without a mesal finger, the arista sub-basal and not apical, and the presence of large conspicuous claspers on the fourth and fifth sternites in contrast to the inconspicuous clasping organ of the Neriidae.

KEY TO THE SUBFAMILIES AND GENERA OF MICROPEZIDAE

1. Bases of antennae approximate; male claspers variable, often greatly enlarged; ovipositor variable (figs. 2, 3)....2
   Bases of antennae distinctly separated; male claspers uniform and inconspicuous; ovipositor broad on basal one-third, then tapering to slender apical one-third (fig. 4) (Subfamily Taeniapterinae) ......*Taeniaptera*[1] Macquart
2. Second basal crossvein present (fig. 5); hind tibia without extensor setae; head approximately as long as wide (fig. 6) (Subfamily Calobatinae) 3
   Second basal crossvein absent; hind tibia with extensor setae; head approximately 1.5 × as long as wide (fig. 24) (Subfamily Micropezinae) ....*Micropeza* Meigen
3. Maxillary palpus short, ratio of length to width, 1:1; pterostigma small or absent; claspers turgid, directed posteriorly, fulcrum situated between their bases (fig. 16)....
   Compsobata Czerny
   Maxillary palpus long, ratio of length to width, 4:1; pterostigma well developed, as long as posterior crossvein (fig. 5); claspers filiform, directed anteriorly;

fulcrum situated on metasternum at bases of hind coxae
(fig. 14) ....*Cnodacophora* Czerny

Subfamily Calobatinae

Calobatinae Enderlein (in part), 1922, Arch. f. Naturg., Abt. A,
88(5):141.

The characters given in the key to subfamilies and
genera will easily differentiate the Calobatinae from
other subfamilies. Additional characters of importance
are: frons rather uniform in sculpturing without well
marked areas, usually entirely opaque; parafrons and
mesofrons scarcely differentiated; sternopleural comb
with one or two hairs or bristles; claspers well developed
in all Nearctic species; fulcrum present in all species
except *Cnodacophora nasoni* (Cresson); and ovipositor
generally robust. A more detailed description of subfamily characteristics and synonymy is given in Cresson
(1938*a*).

Genus *Cnodacophora* Czerny

Cnodacophora Czerny, 1930, *in* Lindner, Die Fliegen der Palae-
arktischen Region, fam. 42:4.

This genus is easily distinguished from *Compsobata*
by the long maxillary palpi and the well-developed
pterostigma in the wing (fig. 5). The claspers are filiform, curving anteriorly (fig. 14). There is only one
species in western North America.

Cnodacophora nasoni (Cresson)
(Figs. 5, 14)

Calobata nasoni Cresson, 1914, Entomol. News, 25:459. Type ♂,
Algonquin, Illinois (Acad. Nat. Sci. Phila., no. 6078).

Geographic range. — Alaska to Quebec, south to Oregon, Colorado, Illinois, and New York.

California records. — None.

Dicussion. — Since the genus *Cnodacophora* consists
of only one species we have included it in the key to
subfamilies and genera of Micropezidae. Females of *C.
nasoni* are sometimes confused with those of *Compsobata mima* because of the similar coloration but the long
white maxillary palpi and well-defined pterostigma
(fig. 5) will easily differentiate the two. The males
should offer no problems.

Genus *Compsobata* Czerny

Compsobata, Czerny, 1930, *in* Lindner, Die Fliegen der Palaerk-
tischen Region, fam. 42:5.

The important characteristics of this genus are given
in the key to subfamilies and genera.

KEY TO THE SUBGENERA AND SPECIES OF GENUS COMPSOBATA

1. Ferruginous species; fulcrum of claspers absent (figs. 8,
   9); ovipositor compressed laterally at base (Subgenus
   Compsobata) ....*univitta* (Walker)
   Black species; fulcrum present (fig. 16); ovipositor not
   compressed laterally at base (Subgenus *Trilophyrobata*)..2
2. Claspers large, about equal in size to globose postabdomen
   (fig. 18); ovipositor broad, basal angles strongly auriculate with lobes building outward (fig. 10)...*pallipes* (Say)
   Claspers smaller than globose postabdomen (fig. 12); ovipositor usually subcylindrical, not auriculate basally or
   feebly so with lobes less prominent....................3
3. Fulcrum U-shaped with two distinct lobes (fig. 15); meso-
   pleuron AEPS$_2$) dull, cinereous-pollinose....*jamesi* Merritt
   Fulcrum not U-shaped and without distinct lobes; meso-
   pleuron (AEPS$_2$) generally shining 4
4. Fulcrum heart-shaped (figs. 12, 13, and 16), its maximum
   width 0.3 mm; ovipositor subcylindrical, not auriculate
   basally (fig. 11); femora usually pale but often with a
   dark area dorsally ....*mima* (Hennig)
   Fulcrum weakly bilobed (fig. 17), maximum width less than
   0.2 mm; ovipositor not subcylindrical and feebly auriculate basally; femora predominantly dark ....
   ....*microfulcrum* (James)

Subgenus *Compsobata* Czerny

Subgeneric characters are given in the key to the
genus *Compsobata.*

Compsobata (Compsobata) univitta (Walker)
(Figs. 8, 9; map 2)

Calobata univitta Walker, 1849, List Insects British Museum,
Dipt., 4:1049. Type ?, New York, and St. Martin's Falls, Al-
bany River, Hudson Bay [British Museum (Natural History)].

Calobata univittata Johnson, 1900, in Smith's Ins. N.J., 2:692.
Error or Emend.

Calobata albiceps Wulp, 1883, Tijds. V. Entomol., 26:50.

Calobata agilis Harris, 1835, Hitchcock's Rep. Geol. Bot. Zool.
Mass., 2:600. Nomen nudum.

Geographic range. — British Columbia to Quebec, south to
California, New Mexico, Illinois, Maryland.

---

[1] This genus and subfamily probably do not occur in California.
It is chiefly Neotropical. One species, *T. lasciva* (F.), has been
recorded in the literature as from California, but this is an
error; the locality should be San José del Cabo, Baja California,
Sur.

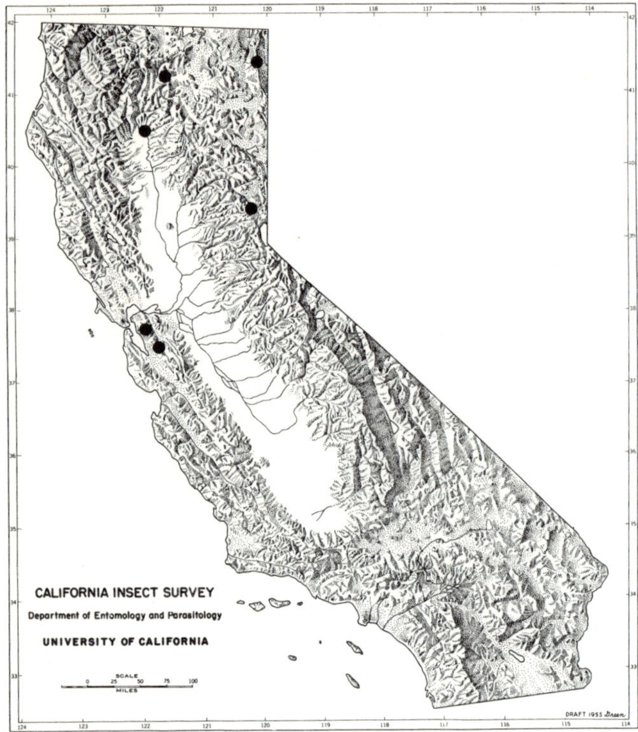

Map 2. California distribution of *Compsobata (Compsobata) univitta* (Walker).

California records.—ALAMEDA Co.: Sunol, ♂, IV-26-39 (M. A. Cazier, A.M.N.H.). CONTRA COSTA Co.: Pinehurst Cyn. Rd., ♂, V-16-71 (E. J. Rogers, Sr., pers. coll.). MODOC Co.: Cedar Pass, ♂, VI-29-55 (E. E. Lindquist, C.I.S.). NEVADA Co.: Hwy. 89, ¼ mi. S. Hobart Mills, shaded marsh, 5,800 ft., ♂, VII-18-67 (R. E. Orth, U.C.R.). SHASTA Co.: Redding, ♀, VI-10-29 (A. L. Melander, A.N.S.P.). SISKIYOU Co: 1 mi. N. W. Bartle, ♀, VII-20-66 (P. Rude, C.I.S.).

Discussion. — This is a distinctive species differing from other North American members of the subfamily in its general ferruginous color, and the absence of a fulcrum (figs. 8 and 9). In the female the ovipositor is compressed laterally at the base. Cole (1969) found this species fairly common in shady, moist areas along the banks of the Hood River in Oregon. It has also been collected from a stream bed habitat in the Pullman, Washington, area.

Subgenus *Trilophyrobata* Hennig

Trepidaria, subg. *Trilophyrobata* Hennig, 1938, Insecta Matsumurana, 13:9.

The subgenus *Trilophyrobata* consists of four species found in western United States and one species, *Calobata kennicotti* (Banks), found in the Hudson Bay Territory and not discussed in this manuscript.

The general black color and presence of a fulcrum (fig. 16F) easily distinguishes this subgenus from the subgenus *Compsobata.*

Compsobata (Trilophyrobata) jamesi Merritt
(Fig. 15)

Compsobata (Trilophyrobata) jamesi Merritt, 1971, Pan-Pac. Entomol., 47:179. Type ♂, Nahcotta, Pacific County, Washington (Washington State University).

Geographic range. — Washington.

California records. — None.

Discussion. — The males of *jamesi* can be separated from other males by the U-shaped fulcrum with two distinct lobes (fig. 15). We examined only one female, the allotype. In this specimen the ovipositor was deformed. It appears to be of the *mima* type, subcylindrical and not auriculate basally. The pollinose mesopleura (AEPS$_2$) will distinguish both sexes of *jamesi* from those of *mima.*

Compsobata (Trilophyrobata) microfulcrum (James)
(Fig. 17)

Paracalobata microfulcrum James, 1946, Entomol. News, 57:129. Type ♂, Divide, Teller County, Colorado (Colorado State University).

Geographic range. — Colorado, Arizona.

California records. — None.

Discussion. — This species, not included in any previous keys, can be distinguished by the small fulcrum (fig. 17) and the extensive darkening of the legs. The ovipositor of microfulcrum is feebly auriculate basally; in *mima* it is completely subcylindrical (fig. 11); in *pallipes* it is strongly auriculate basally (fig. 10). It is difficult to examine ovipositors of pinned specimens because of abdominal shrinkage.

Compsobata (Trilophyrobata) mima (Hennig)
(Figs. 11-13, 16; map 3)

Trepidaria mima Hennig, 1936, Konowia, 15:134. Lectotype ♂ designation by Cresson), Electra Lake, La Platte County, Colorado (Amer. Mus. Nat. Hist.).

Geographic range. — British Columbia to Saskatchewan, south to California and Colorado.

California records. — ALPINE Co.: Woodfords, ♀, VI-17-58 (W. W. Middlekauff, C.I.S.). Woods Lake, ♀, IX-8-63 (I. Savage, C.D.A.). EL DORADO Co.: Lake Tahoe, 4 ♂, VIII-2-40

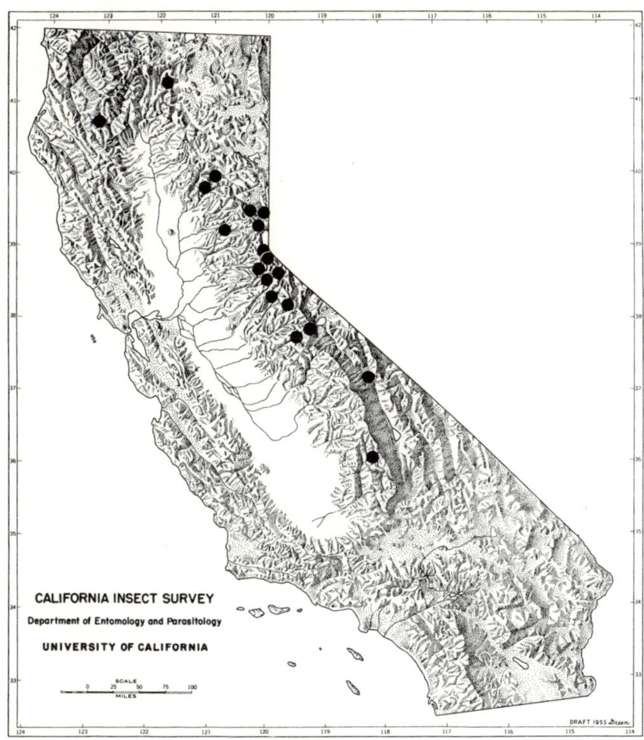

Map 3. California distribution of *Compsobata (Trilophyrobata) mima* (Hennig).

TULARE CO.: Deadman Canyon, Kings Canyon Nat. Park, ♀, 2 ♂, VII-6-61 (R. E. Rice, U.C.D.). Tuolumne Meadows, ♀, 12 ♂, VII-1-40 (D. E. Hardy, U.K.). Dardanelles, ♂, VI-26-51 (A. T. McClay, U.C.D.).

Discussion. — Compsobata mima is the most common species west of the Rocky Mountains. It is usually found in marsh and meadow habitats at altitudes up to 10,000 feet. Merritt (1970) discussed regional melanism involving leg coloration in *Compsobata mima*. Many specimens have a dark brown to black marking on the middorsal aspect of each femur, while in others this marking is only moderately dark brown and in still others it is lacking. Geographical evidence indicates that this variation is environmental and not genetic. Definite trends toward increased melanism were shown in relation to increases in altitude, latitude north, and longitude west. In many specimens this melanic marking on the femora will help differentiate *mima* from *microfulcrum*.

The male fulcrum shows some variation (figs. 12, 13, and 16), but not enough to cause confusion with the smaller and characteristically shaped fulcra of either *microfulcrum* or *jamesi* (figs. -5 and 17). The claspers of *pallipes* (fig. 18) are larger than in *mima*. The ovipositor of *mima* is subcylindrical (fig. 11).

(L. J. Lipovsky, U.K.); ♀, ♂, VII-23-27 (E. H. Nast, C.A.S.); ♀, ♂, VIII-2-40 (D. E. Hardy, U. K.); 2 ♀, VIII-2-40 (F. E. Kenaga, U.K.); ♀, 4 ♂, VI-19-36 (R. M. Bohart, U.C.D.). Bijou, Lake Tahoe, ♀, VI-30-29 (R. L. Usinger, C.A.S.); 2 ♂, VI-23-53 (G. F. Knowlton, U.S.U.). Echo Lake, ♀, VII-17-61 (W. W. Middlekauff, C.I.S.); ♀, VII-21-48 (S. A. Sher, C.I.S.). Luther Pass, ♀, VII-24-55 (E. I. Schlinger, U.C.D.); ♀, VII-8-64 (W. K. Thrailkill, U.I.). INYO CO.: Bishop, 7 ♂, VII-28-40 (D. E. Hardy, U.K.); ♂, VII-28-40 (L. C. Kuitert, U.K.). MONO CO.: Sardine Cr., 2 ♀, 6 ♂, VII-11-51 (A. T. McClay, U.C.D.); ♂, VII-11-51 (R. W. Morgan, C.I.S.); 3 ♀, VII-12-51 (A. T. McClay, U.C.D.). Tioga Pass, ♀, 2 ♂, VII-31-40 (D. E. Hardy, U.K.). Sonora Pass, 2 ♀, VIII-10-60 (E. Jessen, C.I.S.). Nevada U.K.). Sonora Pass, 2 ♀, VIII-10-60 (E. Jessen, C.I.S.). NEVADA CO.: Sagehorn Cr., 2 ♀, ♂, VII-7-64 (M. E. Irwin, U.C.R.); 3 ♂, VII-5-62 (E. J. Montgomery, U.C.D.). Sagehorn Cr. near Hobart Mills, ♀, VII-1-64 (R. E. Scott, U.C.R.); ♀, ♂, VI-25-54 (R. H. Goodwin, C.I.S.). Toll House Lk., ♂, VIII-5-51 (E. I. Schlinger, U.C.D.). Emigrant, ♀, VII-15-51 (E. I. Schlinger, U.C.D.). Truckee, 2 ♀, 3 ♂, VII-6-27 (E. P. Van Duzee, C.A.S.). PLUMAS CO.: Bucks Lake, ♀, 5 ♂, VI-23-49 (W. F. Ehrhardt, U.C.D.); ♀, 2 ♂, VII-14-49 (R. C. Bechtel, U.C.D.). Quincy, 4 mi. W., 2 ♂, VI-30-49 (H. A. Hunt, U.C.D.). SIERRA CO.: Sierraville, 3.5 mi. W., 6 ♀, 2 ♂, VI-6-67 (R. E. Orth, U.C.R.). Sierraville, 4.8 mi. S.E., 8 ♀, VI-17-54 (R. H. Byers and party, U.K.). Independence Lake, ♂, VII-17-54 (R. H. Goodwin, U.C.). Webber Lake, ♀, VIII-5-57 (E. I. Schlinger, U.C.D.); ♀, VII-2-59 (J. N. Linsley, U.C.); 2 ♀, 2 ♂, VIII-5-51 (E. I. Schlinger, U.C.D.); 2 ♀, 2 ♂, IX-4-63 (I. Savage, C.D.A.). SISKIYOU CO.: 1 mi. N.W. Bartle, 5 ♀, 2 ♂, VII-20-66 (P. Rude, C.I.S.. TRINITY CO.: Eagle Cr., 2 ♀, 2 ♂, VI-28-51 (A. T. McClay, C.A.S.).

Compsobata (Trilophyrobata) pallipes (Say)
(Figs. 10, 18)

Calobata pallipes Say, 1828, J. Acad. Nat. Sci. Philadelphia, 3:97. Type, Missouri; lost.

Calobata alesia Walker, 1849, List. Ins. Brit. Mus., Dipt., 4:1048.

Geographic range. — Alaska, Alberta to New Brunswick, south to Nevada and New Jersey.

California records. — None.

Discussion. — The large male claspers (fig. 18) and the strongly auriculate ovipositor (fig. 10) will distinguish *pallipes* from all other species of the subgenus *Trilophyrobata*. Cresson (1938a) states in his key to *pallipes*, "mesonotum more or less finely sculptured, the pollinose vestiture interrupted or narrowly continued across the prescutum hump," as contrasted with "mesonotum smooth, polished, the vestiture broadly continuous across the prescutum hump," for *mima*. We find this to be difficult character to use, especially in *mima*. In addition it is difficult to observe because most specimens are pinned through the mesonotum.

Subfamily Micropezinae

Micropezinae Enderlein, 1922, Arch. f. Naturg., Abt. A, 88(5): 159.

This subfamily is easily distinguished by the proportion of the head which is approximately 1.5 times as long as the width (fig. 24). Other characters of importance are the usually shining frons, without well-marked areas; the fourth and fifth sternite of the male, which is generally modified into a clasping organ (figs. 21 and 23); ovipositor slender tapering towards apex; absence of the second basal crossvein; and presence of extensor setae on the hind tibiae. For a more detailed description and synonymy of the subfamily see Cresson (1938*a*).

Species of this subfamily are known from the Palaearctic, Neotropical and Nearctic regions (Cresson, 1938*a*).

Genus *Micropeza* Meigen

Micropeza Meigen, 1803, Illig. Mag., 2:276.

The important characters of the North American species of this genus are given in the foregoing key and are also discussed under the subfamily heading.

KEY TO THE SUBGENERA AND SPECIES OF THE GENUS MICROPEZA

1. Notopleuron with two bristles (Subgenus *Neriocephalus*)..11
   Notopleuron with one bristle (Subgenus *Micropeza*).....2
2. Thorax black or variable, but with a broad pale notopleural stripe 3
   Thorax black without broad pale notopleural stripe......8
3. Mesonotum noticeably vittate, at least behind suture.....6
   Mesonotum uniformly black 4
4. Parencephalon and occiput pale on lower half (fig. 20); antenna of male pale, of female black; claspers long, almost attaining base of abdomen (fig. 19)........
   ambigua Cresson
   Parencephalon and occiput, except narrow oral margin, dark brown to black (fig. 37); antenna of male and female black; claspers short, not attaining base of abdomen 5
5. Claspers as in fig. 21; sternites of female with strong black marginal setae (fig. 38); inner margin of dark area on epicephalon bent anteriorly at a distinct angle as in fig. 22 ....*setaventris* Cresson
   Claspers as in fig. 23; sternites of female without strong black marginal setae; inner margin of dark area on epicephalon only gently bowed as in fig. 24 [(California)] ....*unca* Merritt
6. Dark horseshoe shaped marking on vertex and occiput uniformly broad throughout (fig. 26); antenna black; vertex broadly black laterally (fig. 26); mesonotum pale with three black stripes (fig. 36); claspers long, almost attaining base of abdomen (fig. 25)........
   turcana Townsend
   Dark horseshoe shaped marking on vertex and occiput consisting of a thin line or with a broad base but never remaining broad throughout 7

7. Claspers short, extending slightly basad of apex of tergite II (fig. 28); male with posterior surface of hind femur densely pilose; dark pigmentation on paracephalon of female extending below outer vertical bristle approximately to middle of eye (fig. 29).....*lineata* Van Duzee
   Claspers long, almost attaining base of abdomen (fig. 30); posterior surface of hind femur sparsely pilose; dark pigmentation of parencephalon rarely extending below outer vertical bristle and never to middle of eye (fig. 31) ....*compar* Cresson
8. Thorax shining; claspers as in fig. 41 ....*nitidor* Cresson
   Thorax dull, pollinose 9
9. Antennae pale, arista entirely white;[2] femora with dark distomedian and apical flexor spots; claspers as in fig. 7; postabdomen with a black circular spot.........
   abnormis Cresson
   Antennae black to dark brown, arista black basally, apical portion variable; femora without spots; postabdomen without circular spot 10
10. First posterior cell open; claspers dark, each terminating in a sternal prong (fig. 27); ovipositor tapering toward apex, pointed at tip (fig. 2)................*atra* Cresson
    First posterior cell barely closed; claspers lacking sternal prongs, with simple apical expanded plate (figs. 32 and 33); ovipositor tapering towards apex, but truncate at tip (fig. 34) ....*ventralis* Cresson
11. Arista black; dark densely pollinose species with two shining lines on vertex between inner and outer vertical bristles ....*bisetosa* Coquillett
    Arista white; shining lines on vertex absent.............12
12. Each femur with a distomedian and a subapical black ring; thorax with brown markings (fig. 39); claspers as in fig. 40 ....*stigmatica* Wulp
    Each femur without black rings; thorax without brown markings 13
13. Mesonotum uniformly cinereous or ferruginous with or without longitudinal markings; femora without dark proximo-median ring; claspers and styli as in figs. 42 and 43 ....*ruficeps* Wulp
    Mesonotum black except for pale reddish-yellow lateral margins which are broader in presutural area; each femur with a dark but inconspicuous proximo-median ring (males unknown)...........*californica* Van Duzee

Subgenus *Micropeza* Meigen

The presence of only one notopleural bristle will distinguish the subgenus *Micropeza* from subgenus *Neriocephalus*. In addition, the first posterior cell may be open or closed at the margin.

Micropeza (Micropeza) abnormis Cresson
(Fig. 7)

Micropeza abnormis Cresson, 1938, Entomol. News, 49:72. Type ♂, Baboquivari Mountains, Pima County, Arizona (Univ. Kans.).

------

[2] Observe against black background.

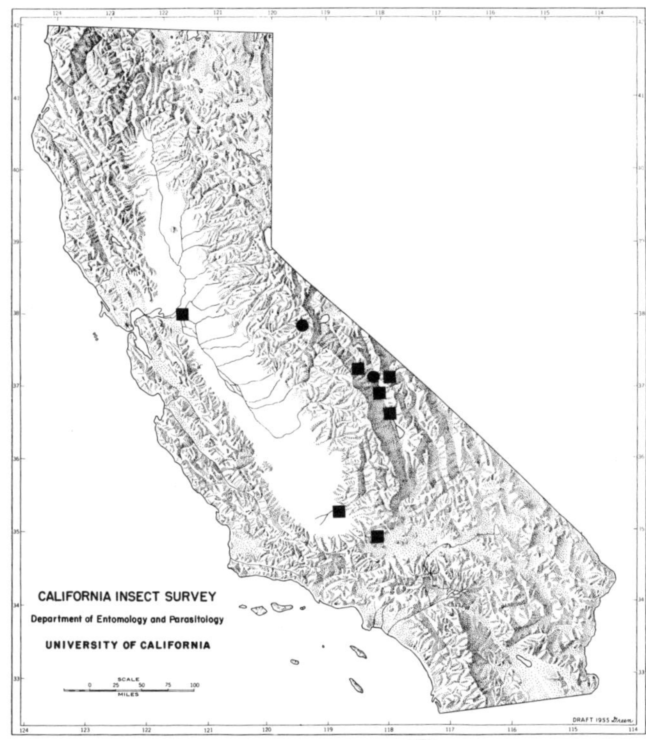

Map 4. California distribution of: ● *Micropeza (Micropeza) ambigua* Cresson; ■ ,*Micropeza (Micropeza) lineata* Van Duzee.

Geographic range. — Known only from Arizona.

California records. — None.

Discussion. — This species differs from others in the genus *Micropeza* by having a black thorax and a white arista. Cresson (1938a) describes the claspers as "subglobose, situated close to genitalia segment." The claspers are situated close to the genitalia, but the word "subglobose" does not define the shape of the claspers (fig. 7). A black circular marking on the postabdomen will aid in identification of the males.

Micropeza (Micropeza) ambigua Cresson
(Figs. 19, 20; map 4)

Micropeza turcana ambigua Cresson, 1908, Trans. Amer. Entomol. Soc., 34:3. Lectotype ♂ (designation by Cresson) Cloudcroft, New Mexico (Acad. Nat. Sci. Phila., no. 6011).

Geographic range. — New Mexico, Arizona, California.

California records. — Inyo Co.: Bishop, ♀, VI-20-29 (R. L. Usinger, C.A.S.). Tuolumne Co.: Yosemite National Park, 2♀, ♂, VIII-1-40 (E. E. Kenaga, U.K.).

Discussion. — *Micropeza ambigua* should not be confused with any other species except possibly *lineata*. The claspers of *ambigua* (fig. 19) are considerably larg-

er than those of *lineata* (fig. 28) and the mesonotum of *ambigua* is black as compared to pale vittate in *lineata*. The maxillary palpi are pale. Cresson (1938a) observed long pile on the posterior surface of the hind femur of the male *ambigua*, but we failed to see these, even in the paratypes.

Micropeza (Micropeza) atra Cresson
(Fig. 27)

Micropeza atra Cresson, 1938, Entomol, News, 49:74. Type ♀ Flagstaff, Coconino County, Arizona (U.S. Nat. Mus., no. 27059).

Geographic range. — Arizona, Utah, New Mexico.

California records. — None.

Discussion. — The male of this species was first described by Merritt (1971a). Contrary to the speculation of Cresson (1938a), the claspers are not of the *lineata* type but rather of the *compar-turcana-ambigua* type (fig. 27). The open first posterior cell, large dark claspers, and the tapering, apically pointed ovipositor (fig. 2) will differentiate *atra* from any other species.

Micropeza (Micropeza) compar Cresson
(Figs. 30 and 31)

Micropeza compar Cresson, 1938, Entomol. News, 49:73. Type ♂, Huachuca Mountains, Arizona (Univ. Kans.).

Geographical range. — Arizona, where it is widely distributed, and New Mexico.

California records. — None.

Discussion. — The females of this species are often confused with *lineata* females. Cresson (1938a) separated the two on the basis of the femora and third antennal segment. He stated that *compar* possessed a more or less distinct distomedian extensor spot and had the third antennal segment entirely pale, in contrast to the femora of *lineata* which lacked such a spot and had the third antennal segment pale apically or entirely ferruginous. After examining many specimens, including a paratype, we found that not all specimens of *compar* possess an extensor spot and the antennae of *compar* and *lineata*, in numerous cases, have either faded or become darker with age. Therefore, we have not included either of these characters in our key. In *compar* the dark pigmentation of the female parencephalon rarely extends below the outer vertical bristle and never to the middle of the eye (fig. 31). This holds true in all specimens examined. *Micropeza compar* males are easily distinguished from *lineata* males by the large claspers (fig. 30) and by the absence of pile on the hind femur.

M. compar can be distinguished from *turcana* by the dark horseshoe-shaped marking on the vertex and occiput usually consisting of a thin line and not uniformly broad as in *turcana* (fig. 26).

M. compar is partially sympatric with *lineata* in the region of Mingus Mountain, Yavapai County, Arizona.

This species has been taken on *Apocynum sp.* (Indian hemp) and *Robinia neomexicana* (locust).

Micropeza (Micropeza) lineata Van Duzee
(Figs. 28, 29, 35; map 4)

Micropeza lineata Van Duzee, 1926, Pan-Pac. Entomol., 3:2. Type ♂, Utah Lake, Lehi, Utah County, Utah (Calif. Acad. Sci., no. 1879).

Geographic range. — Alberta to Manitoba, south to California, New Mexico and Nebraska.

California records. — Inyo Co.: Lone Pine, ♀, ♂, VI-14-37 (J. W. Johnson, A.N.S.P.); 2♀, VI-9-29 (R. L. Usinger, A.N.S.P.); 2♀, VI-14-37 (J. H. Mitchell, A.M.N.H.); ♂, VII-28-40 (L. J. Lipovsky, U.K.). Big Pine, 6♀, 2♂, VI-17-29 (R. L. Usinger, A.N.S.P.), 10 mi. N.W. of Bishop, 2♀, ♂, VI-30-61 (D. R. Miller, U.C.D.). Fish Slough, 10 mi. E of Bishop, ♀, VI-5-67 (J C. Hall, U.C.R.). Kern Co.: Bakersfield, 3♀, VI-4-29 (E. P. Van Duzee, A.N.S.P.). Tehachapai Pass, 2♀, VI-6-29 (R. L. Usinger, A.N.S.P.). Sacramento Co.: Isleton, on field corn, 5♀, VIII-5-65 (K. Miller, C.D.A.).

Discussion.—This species has the largest range of any western Nearctic member in the genus. It also has the greatest degree of color variation. The vertex of *lineata* is usually pale with linear markings (fig. 35), while the mesonotum is generally vittate (pale with three black stripes or dark with two pale stripes) but often the stripes are barely distinguishable or the mesonotum is broadly reddish.

There are two forms of *lineata* females, one with black antennae and one with pale to brown antennae. The presence, in *lineata*, of a dark pigmentation on the female parencephalon extending below the outer vertical bristle and approximately to the middle of the eye will serve to distinguish the females from *compar* females. The claspers and, to a lesser extent, pilosity of the hind femur will separate males of both species.

The dark horseshoe-shaped marking on the vertex and occiput of *lineata* is never uniformly broad throughout as in *turcana*. At most it may be broad at the base but never remaining so throughout. The vertex of *lineata* is never broadly black laterally and the mesonotal pattern of stripes is not distinct as in *turcana*.

M. lineata has been taken on field corn in California.

Micropeza (Micropeza) nitidor Cresson
(Fig. 41)

Micropeza nitidor Cresson, 1936, Trans. Amer. Entomol. Soc., 44:319. Type ♀, Bear Wallow, Santa Catalina Mts., Arizona (Amer. Mus. Nat. Hist.).

Tylos nitidus Hennig, 1936, Konowia, 15:212.

Geographic range. — Known only from mountains of southern Arizona.

California records. — None.

Discussion.—This species is easily distinguished from all others by the glossy black thorax and the absence of a notopleural stripe.

Micropeza (Micropeza) setaventris Cresson
(Figs. 21, 22, 37)

Micropeza setaventris Cresson, 1936, Entomol. News, 49:74. Type ♂, Fort Duchesne, Utah (Acad. Nat. Sci. Phila., no. 6536).

Geographic range. — Utah, Colorado and North Dakota, south to Arizona and New Mexico.

California records. — None.

Discussion. — This species differs from *lineata* in having a black mesonotum without any vittate pattern and in having the yellow part of the body more brownish-yellow. It is distinguishable from *unca* by the claspers (fig. 21), the strong marginal setae on the female sternites (fig. 38), and the inner margin of the dark area on the epicephalon being bent anteriorly at a distinct angle (fig. 22). It has been recorded from California, but all specimens, determined as this species, that we have examined from California are referred to the recently described species *unca*.

Micropeza (Micropeza) turcana Townsend
(Figs. 25, 26, 36)

Micropeza turcana Townsend, 1892, Trans. Kans. Acad. Sci., 13:136. Lectotype ♂, (designation by Cresson) Turkey Tanks, Coconino County, Arizona (Univ. Kans.).

Micropeza jamesi Cresson, 1935, Entomol. News, 46:229.

Geographic range.—Alberta to Manitoba, south to Arizona and Kansas.

California records. — None.

Discussion. — *M. turcana* can be distinguished from *lineata* and *compar* by the uniformly broad, dark horseshoe-shaped marking on the vertex and occiput (fig. 26). Other characters of turcana are: black antennae of males and females; vertex broadly black laterally; and mesonotum pale with three distinct black stripes (fig. 36).

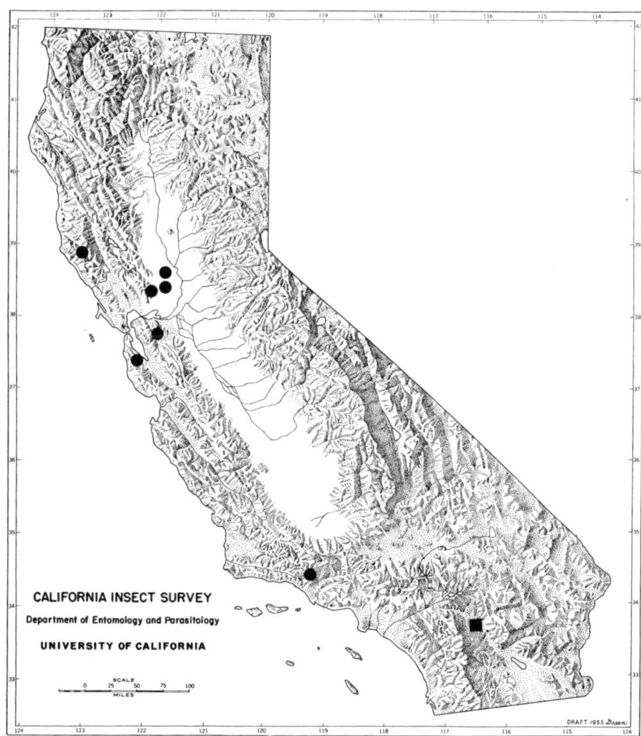

Map 5. California distribution of: ⬤ , *Micropeza (Micropeza) unca* Merritt; ⬛ , *Micropeza (Neriocephalus) californica* Van Duzee.

This species has been taken on *Cleome serrulata* (Rocky Mountain bee plant).

Micropeza (Micropeza) unca Merritt
(Figs. 23, 24; map 5)

Micropeza (Micropeza) unca Merritt, 1971, Pan-Pac. Entomol., 47:181. Type ♂, Davis, Yolo County, California (University of California, Davis).

California records. — CONTRA COSTA CO.: Mt. Diablo, 5♀, ♂, VII-1937 (M. A. Cazier, A.M.N.H.). MENDOCINO CO.: Yorkville, ♂, ♀, IV-24-28 (E. P. Van Duzee, C.A.S.). SACRAMENTO CO.: Sacramento, ♀, collected from *Salix* sp., V-21-33 (H. H. Keifer, C.D.A.). SANTA CLARA CO.: Stanford Univ., 2♂, VI-20-10 (R. W. Doane, C.A.S.). SOLANO CO.: Putah Cyn., ♀, VI-6-48 E. I. Schlinger, U.C.D.). VENTURA CO.: Sespe Cyn., 3♀, VII-10-59 (R. M. Bohart, U.C.D.); ♂, (C. A. Campbell, C.I.S.); ♀, ♂, (P. E. Paige, U.C.D.); ♀, (F. D. Parker, U.C.D.); ♀, (R. W. Spore, U.C.D.). YOLO CO.: Davis, 9♀, 3♂, IV-23-53 (J. C. Hall, U.C.D.); 2♀, ♂, V-14-56 (C. R. Kovacic, U.C.D.). Woodland, ♀, ♂, V-25-59 (F. D. Parker, U.C.D.).

Discussion. — Cresson (1938*a*) apparently failed to note the difference in male claspers in specimens of *setaventris* from California. There is a similarity but upon close examination one can see that the claspers of *unca* (fig. 23) are remarkably different from those of *setaventris* (fig. 21). Also, the inner margin of the dark area on the epicephalon is only gently bowed in *unca* (fig. 24) as opposed to being bent anteriorly at a distinct acute angle in *setaventris* (fig. 22). Also, the sternites of the female do not possess strong black marginal setae.

Micropeza (Micropeza) ventralis Cresson
(Figs. 32-34)

Micropeza ventralis Cresson, 1930, Trans. Amer. Entomol. Soc., 56:356. Type ♂, Tacubaya, Distrito Federal, Mexico (U. S. Nat. Mus., no. 43149).

Geographic range. — Mexico (D.F.), Arizona.

California records. — None.

Discussion. — This is a dull black species with yellow ventral surfaces and brownish wings. Since we were able to examine only one female, we are greatly indebted to Mr. George Steyskal of the U.S. National Museum for examining the holotype and allotype of *M. ventralis* and allowing us to use his drawings and comparative analysis of the species. The ovipositors of *atra* and *ventralis* both taper, but that of *atra* is more pointed (fig. 2), while the ovipositor of *ventralis* is truncate with the tip at least one-third the greatest width (fig. 34). Males of *M. ventralis* differ from either *atra* or *abnormis* by the lack of sternal prongs on the claspers and by having a rather simple, apically expanded plate (figs. 32 and 33). The first posterior cell is barely closed at the margin.

Subgenus *Neriocephalus* Enderlein

Neriocephalus Enderlein, 1922, Arch. f. Naturg., Abt. A, 88(5): 160.

The subgenus *Neriocephalus* can easily be distinguished from subgenus *Micropeza* by the presence of two notopleural bristles. Six species have been found in America north of Mexico and four of these extend west of the Rocky Mountains. Not included in this study are *Micropeza texana* Cresson and *Micropeza producta* Walker, from Texas and southeastern United States, respectively.

Micropeza (Neriocephalus) bisetosa Coquillett

Micropeza bisetosa Coquillett, 1902, J. New York Entomol. Soc., 10:177. Type ♂, Prescott, Arizona (U.S. Nat. Mus., no. 6626).

Geographic range. — Arizona, New Mexico.

California records. — None.

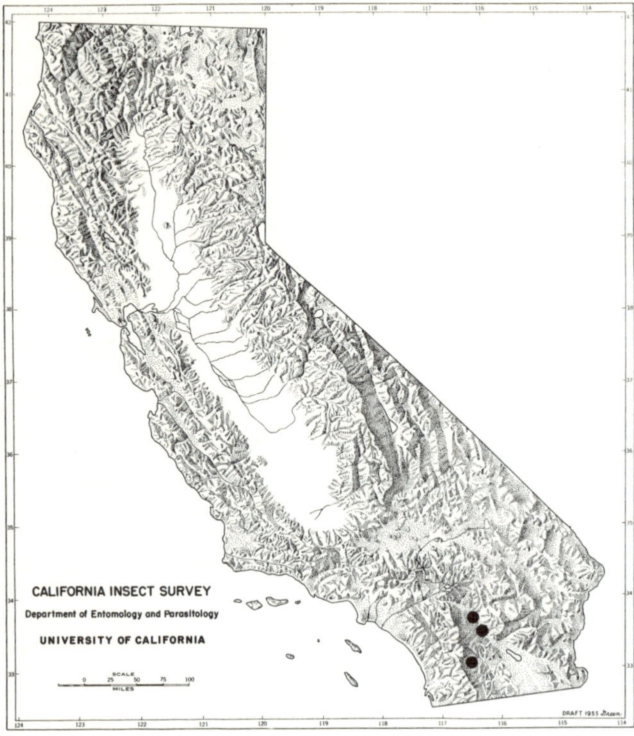

Map 6. California distribution of *Micropeza (Neriocephalus) ruficeps* Wulp.

Discussion.—M. bisetosa can be separated from other members of the subgenus by the presence of a black arista and by the two shining lines on the vertex between the inner and outer vertical bristles.

Micropeza (Neriocephalus) californica Van Duzee
(Map 5)

Micropeza californica Van Duzee, 1926, Pan-Pac. Entomol., 3:1. Type ♀, Palm Springs, Riverside County, California (Calif. Acad. Sci., no. 1878).

Geographic range. — Known only from California.

California records. — RIVERSIDE Co.: Palm Springs, ♀, IV-3-25 (E. C. Van Dyke, C.A.S.; Holotype—no. 1878 C.A.S.).

Discussion. — Through the kindness of Dr. Paul H. Arnaud, Jr., of the California Academy of Sciences, we were able to examine the holotype of *Micropeza californica.* Cresson (1938a) did not examine the type but on the basis of the original description placed it in *Neriocephalus.* The presence of two notopleural bristles on the type verifies its position in this subgenus. *M. californica* is similar to *ruficeps* in regard to coloration of the head but is distinguished from it by the predominantly black mesonotum. The first posterior cell is closed and each femur possesses a dark but inconspicuous proximo-median ring. The males of this species are unknown.

Micropeza (Neriocephalus) ruficeps Wulp
(Figs. 42, 43; map 6)

Micropeza ruficeps Wulp, 1897, Biologia Centrali-Americana, Dipt. 2:365. Type ♀, Northern Sonora, Mexico [British Museum (Natural History)].

Micropeza flaviventris Cole, 1923, Proc. Cal. Acad. Sci., 12:477.

Geographic range. — California, Arizona, Mexico, (Sonora, Baja California).

California records. — RIVERSIDE Co.: Palm Springs, ♂, VII-13-97 (A.N.S.P.). Magnesia Cyn., ♀, IV-21-51 (E. J. Taylor, U.C.D.). SAN DIEGO Co.: Sentenac Cyn., ♀, IV-23-51 (E. I. Schlinger, U.C.D.).

Discussion. — There appear to be two forms of *M. ruficeps,* one in which the mesonotum is uniformly cinereous with two distinct brownish stripes and the other where the mesonotum is ferruginous with few longitudinal markings. The claspers (fig. 42 and 43) are the same in both forms. All forms of *ruficeps* may be distinguished from *stigmatica* by the absence of the black rings on the femora and from *californica* by the absence of a black thorax.

This species has been taken on *Franseria deltoidea* in Arizona.

Micropeza (Neriocephalus) stigmatica Wulp
(Figs. 39 and 40)

Micropeza stigmatica Wulp, 1897, Biologia Centrali-Americana, Dipt. 2:366. Cotypes, Mexico, several localities [British Museum (Natural History)].

Geographic range. — Arizona, New Mexico, Kansas, Texas, south to Argentina.

California records. — None, but widely distributed in Arizona and probably extending into California.

Discussion. — This species differs from all others in having two distinct black rings on each femur, one distomedian and one subapical. The characteristic brown markings of the thorax (fig. 39) and the shape of the claspers (fig. 40) will also aid in identification.

It has been taken on cotton and alfalfa in Arizona.

LITERATURE CITED

Berg, C. O.
1947. Biology and metamorphosis of some Solomon Islands Diptera. Part I: Micropezidae and Neriidae. Occas. Pap. Mus. Zool. Univ. Michigan No. 503:1-14.

Brindle, A.
1965. Taxonomic notes on the larvae of British Diptera. No. 19, The Micropezidae (Tylidae). The Entomologist, 98 (1233):83-86.

Cole, F. R.
1927. A study of the terminal abdominal structures of male Diptera (two-winged flies). Calif. Acad. Sci. Proc. Ser. 4, 16:397-499.
1969. The Flies of Western North America. University of California Press, Berkeley. 693 pp.

Crampton, G. C., C. H. Curran, and C. P. Alexander
1942. Guide to the Insects of Connecticut. Part VI. The Diptera or True Flies of Connecticut. First fascicle [part]. The external morphology of the Diptera. Conn. State Geol. and Nat. Hist. Surv. Bull. 64:1-509.

Cresson, E. T., Jr.
1930. Notes on and descriptions of some neotropical Neriidae and Micropezidae (Diptera). Trans. Amer. Entomol. Soc., 56:307-62.
1938a. The Neriidae and Micropezidae of America north of Mexico (Diptera). Trans. Amer. Entomol. Soc., 64:293-366.
1938b. Descriptions of some North American Micropezidae (Diptera). Entomol. News, 49:72-76.

Hennig, W.
1934. Revision der Tyliden (Dipt., Acalypt.). I. Teil: Die Taeniapterinae Amerikas. Stettin. Entomol. Ztg., 95:65-108, 294-330.
1935a. Revision der Tyliden (Dipt., Acalypt.). I. Teil: Die Taeniapterinae Amerikas [concl.]. Stettin. Entomol. Ztg., 96:27-67.
1935b. Revision der Tyliden (Dipt., Acalypt.). II. Teil: Die ausseramerikanischen Taeniapterinae, die Trepidariinae und Tylinae. Konowia, 14:68-92, 192-216, 289-310.
1958. Die Familien der Diptera Schizophora und ihre phylogenetischen Verwandtschaftsbeziehungen. Beitr. Entomol., 8:505-688.

James, M. T.
1946. The dipterous family Tylidae (Micropezidae) in Colorado. Entomol. News, 57:128-131.

Merritt, R. W.
1970. Melanistic variation in *Compsobata mima* (Diptera: Micropezidae). J. Kans. Entomol. Soc., 43:451-455.
1971. New and little known Micropezidae from the western United States (Diptera). Pan-Pac. Entomol., 47:179-183.
1972. Geographic distribution of Micropezidae in the western United States (Diptera). Northwest Sci. 46(1):40–43.

Oldroyd, H. O.
1964. The Natural History of Flies. W. W. Norton, New York. 324 pp.

Sabrosky, C. W.
1942. An unusual rearing of *Rainieria brunneipes* (Cresson) (Diptera: Micropezidae). Entomol. News, 53:283-285.

Steyskal, G. C.
1964. Larvae of Micropezidae (Diptera), including two species that bore in ginger roots. Ann. Entomol. Soc. Amer., 57(3):292-296.
1965. Family Micropezidae, pp. 633-636. *In* A. Stone, et al., eds. A catalog of the Diptera of America north of Mexico. U.S.D.A., Agric. Handbk. 276. 1696 pp.

Teskey, H. J.
1972. The mature larva and pupa of *Compsobata univitta* (Diptera: Micropezidae). Can. Entomol., 104:295-298.

Wallace, J. B.
1969. The mature larva and pupa of *Calobatina geometroides* (Cresson) (Diptera: Micropezidae). Entomol. News, 80:317–321.

Wheeler, W. M.
1924. Courtship of the Calobatas. J. Hered., 15:485-495.

PLATES

FIGURES 1 TO 7

Fig. 1. Micropezid head, generalized. AFr, anterior frontal bristles; Ec, epicephala; F, frontalia; Iv, inner-vertical bristles; Lu, lunula; Mf, mesofrons; Oc, ocelli; Ov, outer-vertical bristles Pc, paracephala; Pf, parafrons; PFr, posterior frontal bristles; Pv, postvertical bristles.

Fig. 2. Subfamily Micropezinae, dorsal view of generalized ovipositor.

Fig. 3. Subfamily Calobatinae, dorsal view of generalized ovipositor.

Fig. 4. Subfamily Taeniapterinae, dorsal view of generalized ovipositor.

Fig. 5. *Cnodacophora nasoni* (Cresson), right wing. 2b, second basal crossvein; pts, pterostigma.

Fig. 6. Subfamily Calobatinae, generalized head.

Fig. 7. *Micropeza (Micropeza) abnormis* Cresson, ventral view of male genitalia.

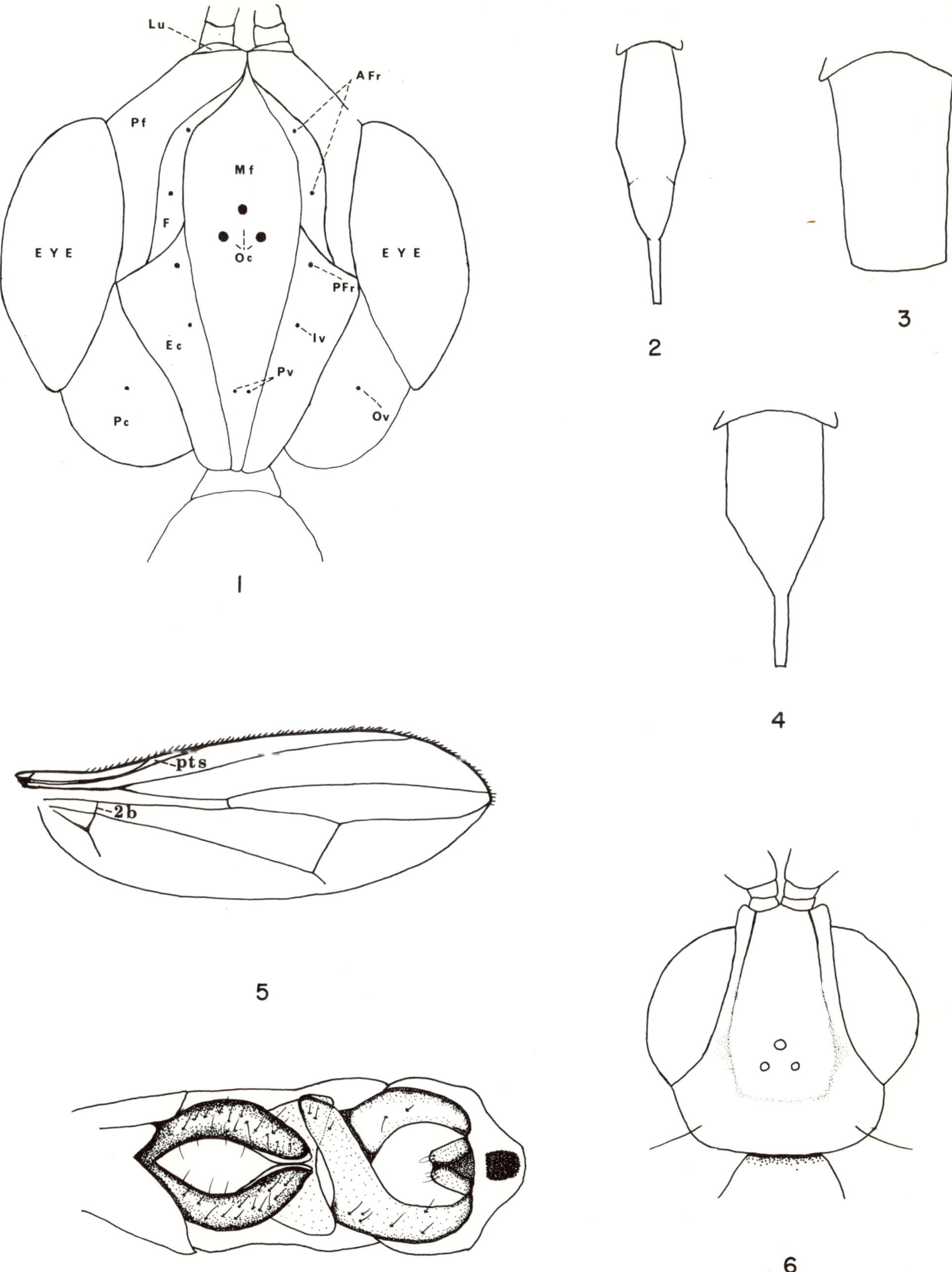

Lu
A Fr
Pf
Mf
F
EYE
Oc
EYE
PFr
Ec
Iv
Pv
Pc
Ov
1
2
3
4
pts
2b
5
6
7

FIGURES 8 TO 14

Fig. 8. *Compsobata (Compsobata) univitta* (Walker), lateral view of apex of male abdomen, showing claspers.

Fig. 9. *Compsobata (Compsobata) univitta* (Walker), ventral view of male abdomen, showing claspers.

Fig. 10. *Compsobata (Trilophyrobata) pallipes* (Say), oblique-lateral view of ovipositor segment.

Fig. 11. *Compsobata (Trilophyrobata) mima* (Hennig), oblique-lateral view of ovipositor segment.

Fig. 12. *Compsbata (Trilophyrobata) mima* (Hennig), lateral view of male abdomen, showing claspers.

Fig. 13. *Compsobata (Trilophyrobata) mima* (Hennig), ventral view of claspers.

Fig. 14. *Cnodacophora nasoni* (Cresson), lateral view of male abdomen, showing claspers.

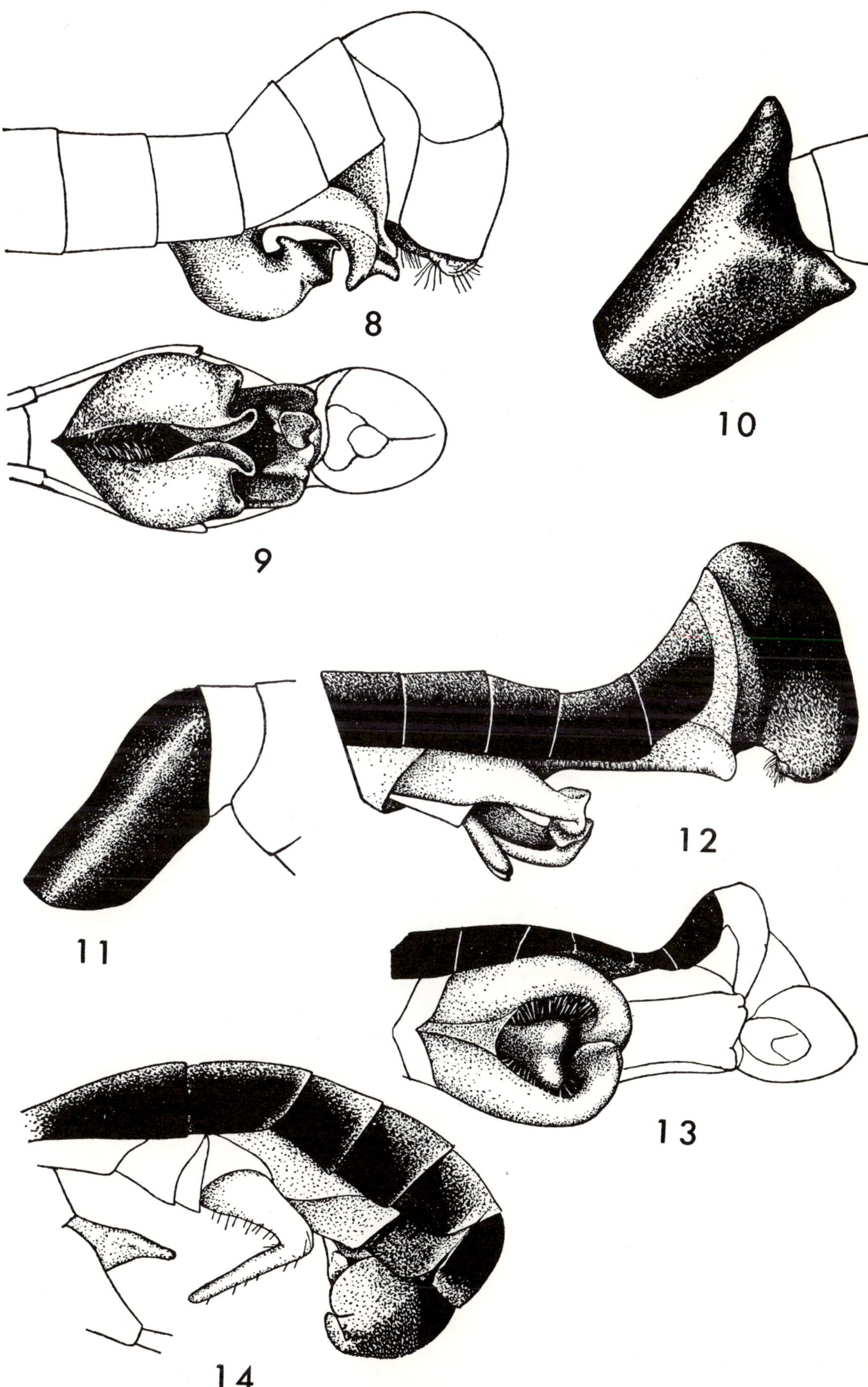

8
9
10
11
12
13
14

FIGURES 15 TO 20

Fig. 15. *Compsobata (Trilophyrobata) jamesi* Merritt, ventral view of claspers.
Fig. 16. *Compsobata (Trilophyrobata) mima* (Hennig), ventral view of claspers.
Fig. 17. *Compsobata (Trilophyrobata) microfulcrum* (James), ventral view of claspers.
Fig. 18. *Compsobata (Trilophyrobata) pallipes* (Say), lateral view of male abdomen, showing claspers.
Fig. 19. *Micropeza (Micropeza) ambigua* Cresson, lateral view of male abdomen, showing claspers.
Fig. 20. *Micropeza (Micropeza) ambigua* Cresson, lateral view of head, showing color pattern.

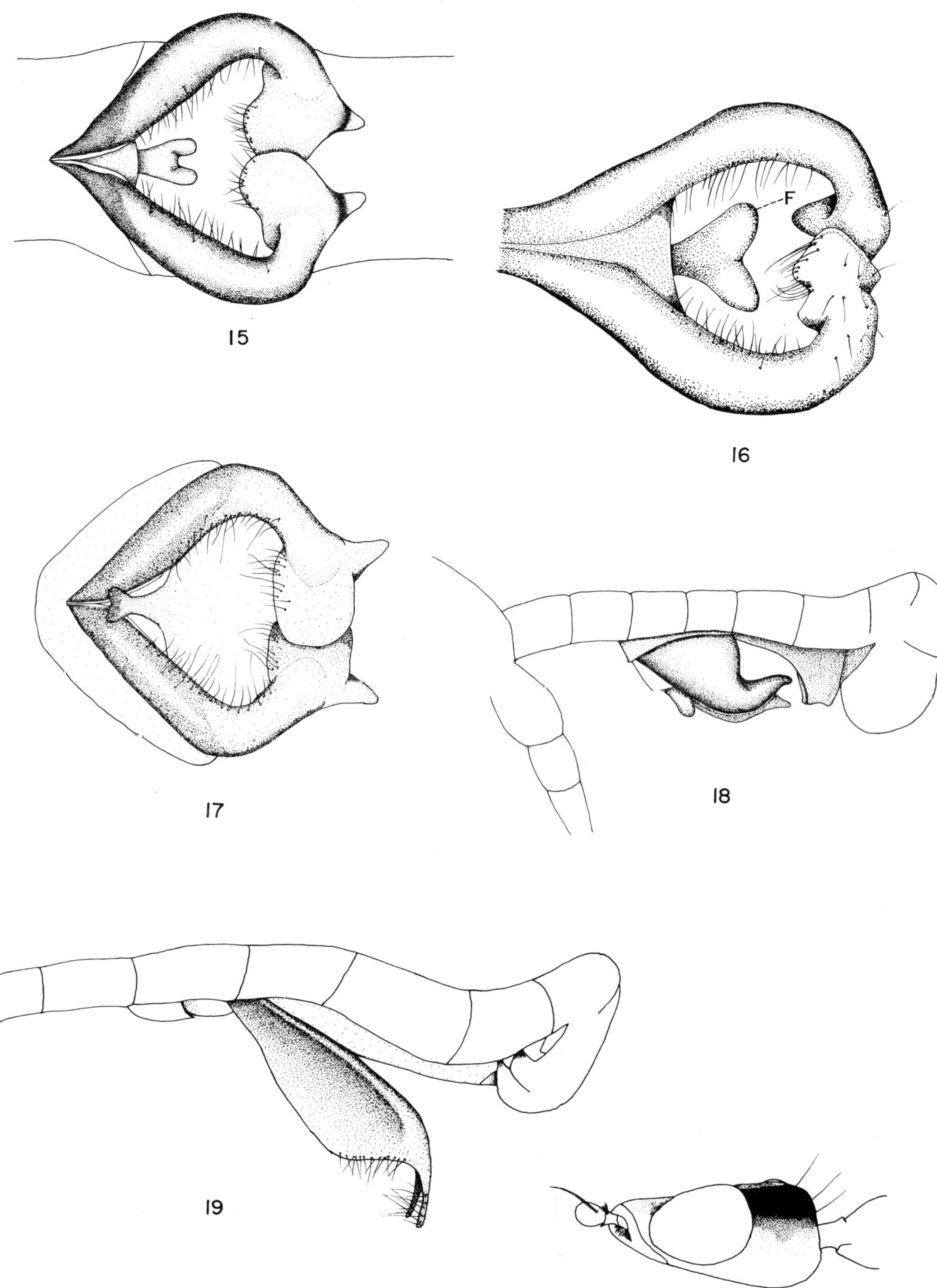

15

16

17

18

19

20

FIGURES 21 TO 27

Fig. 21. *Micropeza (Micropeza) setaventris* Cresson, lateral view of male abdomen, showing claspers.

Fig. 22. *Micropeza (Micropeza) setaventris* Cresson, dorsal view of head, showing color pattern.

Fig. 23. *Micropeza (Micropeza) unca* Merritt, lateral view of male abdomen, showing claspers.

Fig. 24. *Micropeza (Micropeza) unca* Merritt, dorsal view of head, showing color pattern.

Fig. 25. *Micropeza (Micropeza) turcana* Townsend, lateral view of male abdomen, showing claspers.

Fig. 26. *Micropeza (Micropeza) turcana* Townsend, dorsal view of head, showing color pattern.

Fig. 27. *Micropeza (Micropeza) atra* Cresson, lateral view of male abdomen, showing claspers.

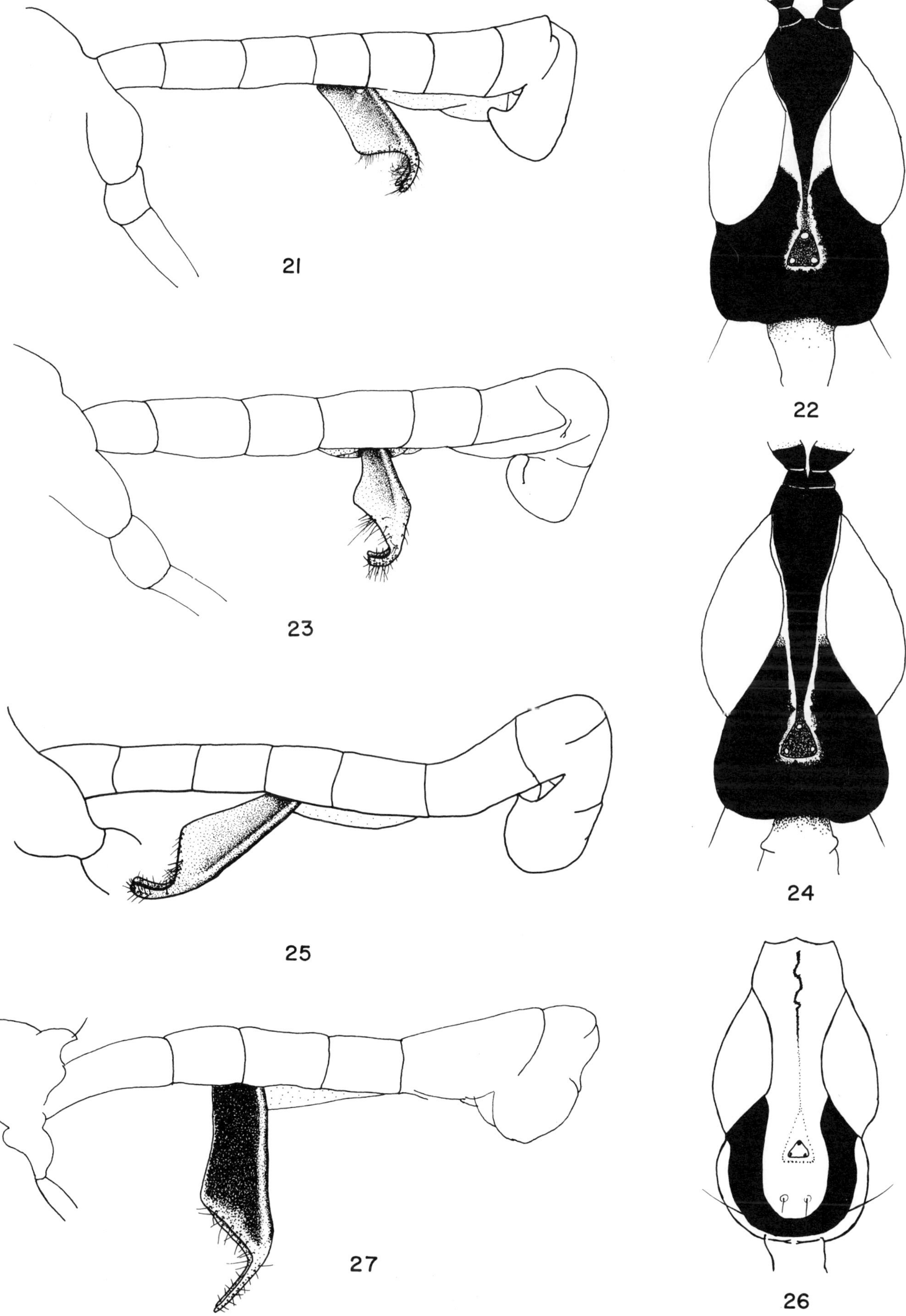

21

22

23

24

25

26

27

FIGURES 28 TO 34

Fig. 28. *Micropeza (Micropeza) lineata* Van Duzee, lateral view of male abdomen, showing claspers.

Fig. 29. *Micropeza (Micropeza) lineata* Van Duzee, lateral view of head, showing color pattern.

Fig. 30. *Micropeza (Micropeza) compar* Cresson, lateral view of male abdomen, showing claspers.

Fig. 31. *Micropeza (Micropeza) compar* Cresson, lateral view of head, showing color pattern.

Fig. 32. *Micropeza (Micropeza) ventralis* Cresson, lateral view of apex of male abdomen, showing genitalia.

Fig. 33. *Micropeza (Micropeza) ventralis* Cresson, ventral view of apex of male abdomen, showing genitalia.

Fig. 34. *Micropeza (Micropeza) ventralis* Cresson, dorsal view of ovipositor.

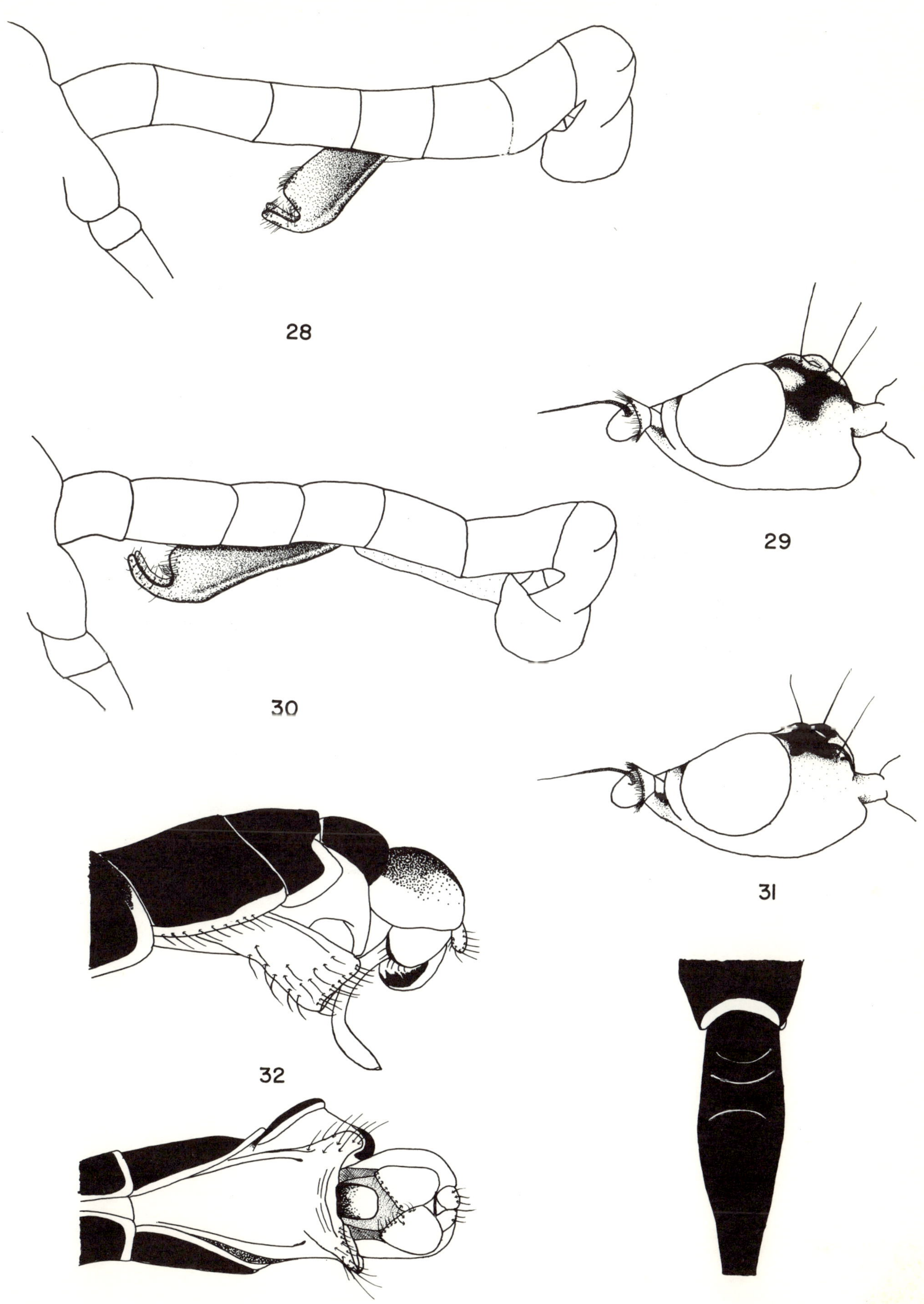

28

29

30

31

32

33

34

FIGURES 35 TO 43

Fig. 35.   *Micropeza (Micropeza) lineata* Van Duzee, dorsal view of head, showing color pattern.

Fig. 36.   *Micropeza (Micropeza) turcana* Townsend, dorsal view of thorax, showing color pattern.

Fig. 37.   *Micropeza (Micropeza) setaventris* Cresson, lateral view of head, showing color pattern.

Fig. 38.   *Micropeza (Micropeza) setaventris* Cresson, lateral view of female abdomen, showing setation of sternites.

Fig. 39.   *Micropeza (Neriocephalus) stigmatica* Wulp, dorsal view of thorax, showing color pattern.

Fig. 40.   *Micropeza (Neriocephalus) stigmatica* Wulp, lateral view of male abdomen, showing claspers.

Fig. 41.   *Micropeza (Micropeza) nitidor* Cresson, lateral view of male abdomen, showing claspers.

Fig. 42.   *Micropeza (Neriocephalus) ruficeps* Wulp, ventral view of male postabdomen, showing styli.

Fig. 43.   *Micropeza (Neriocephalus) ruficeps* Wulp, lateral view of male abdomen, showing claspers.

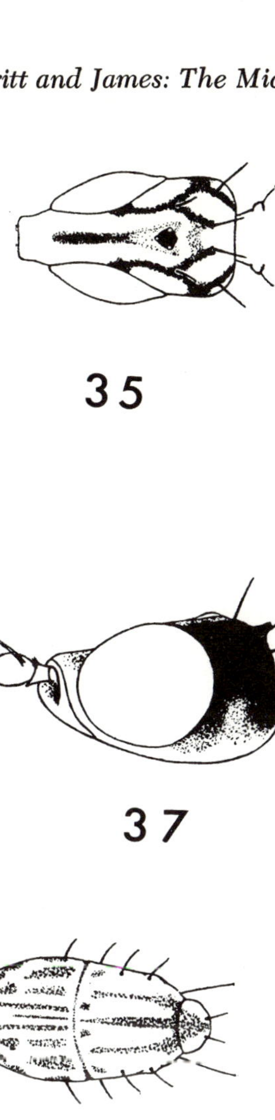

35

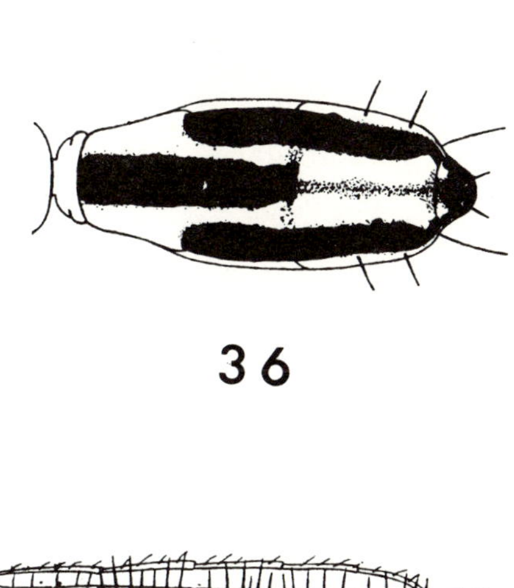

36

37

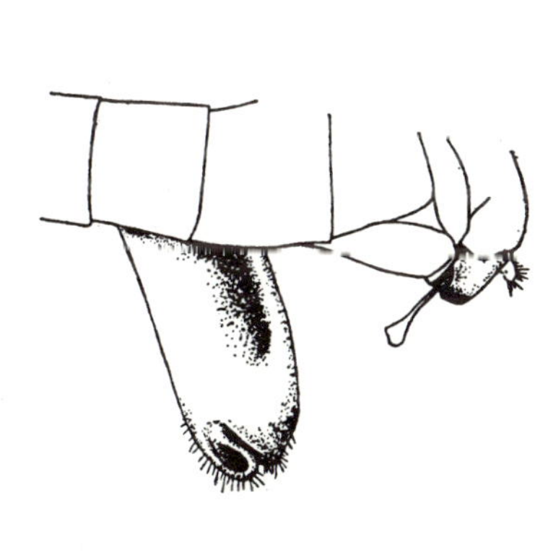

38

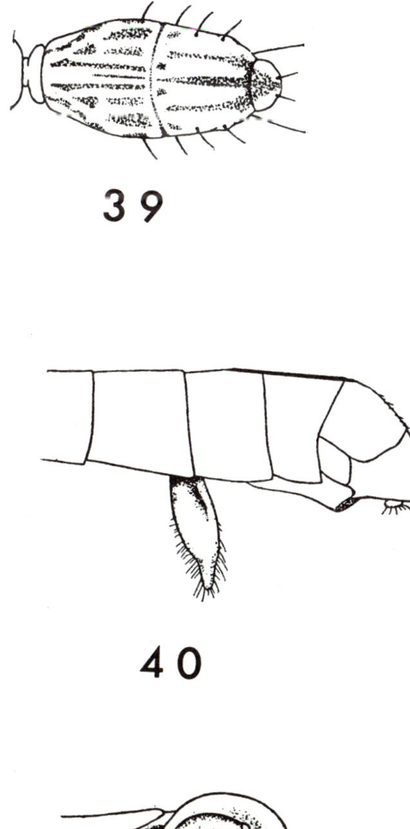

39

41

40

42

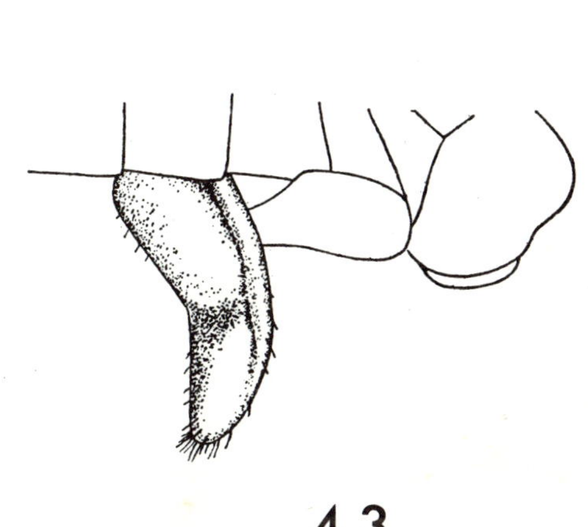

43

INDEX TO THE MICROPEZIDAE OF CALIFORNIA

(Synonyms are in italic; main page references in boldface)